Terrible Maps

Terrible Maps

Hilarious Maps for a Ridiculous World

Michael Howe

HarperCollins*Publishers*

Terrible Maps has been the home of maps with a pinch of humour for nearly ten years. You can find us here:

🐦 @TerribleMaps
📷 @TerribleMap
📘 Terrible Maps

How to find your bearings in Antarctica

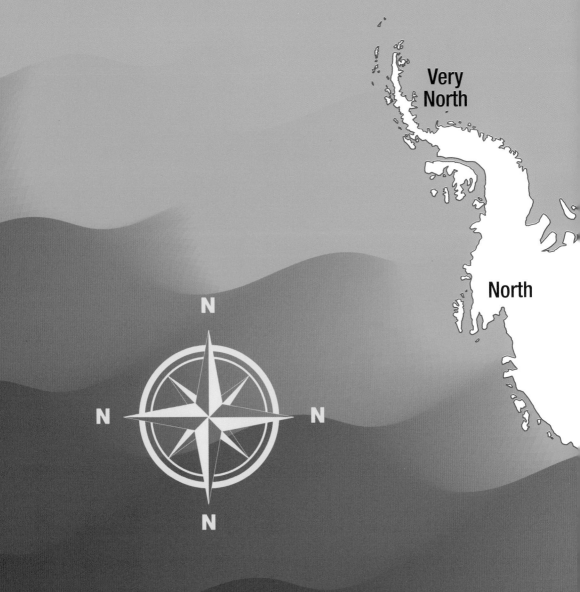

Very
North

North

N

N   N

N

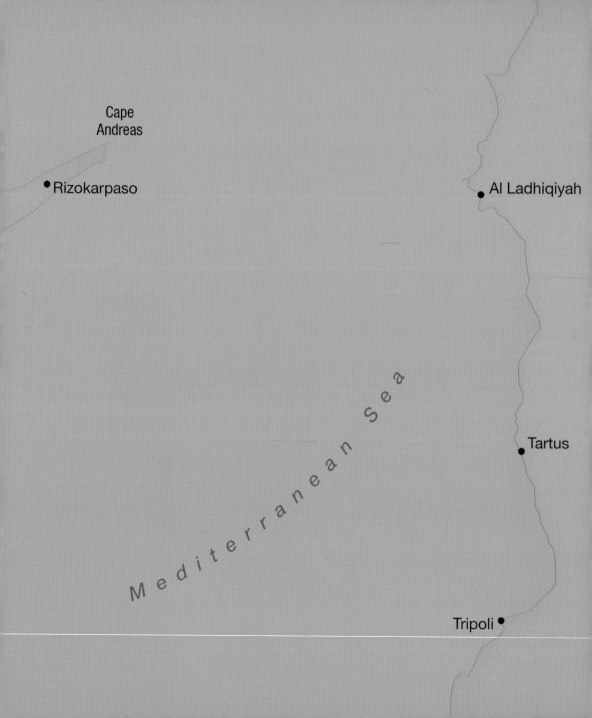

Introduction

Welcome to the world of Terrible Maps, where geographical accuracy takes a backseat and getting lost is practically guaranteed. This book is a collection of maps ranging from humorously inaccurate to completely absurd, where stereotypes are exposed and quirks are celebrated. If you're looking for highly accurate, politically correct and meticulously researched cartography, then you've come to the wrong place. However, if you're ready to embrace the unconventional and indulge in the amusingly terrible side of mapping then keep reading. Whether you are a geography enthusiast, lover of satire or simply looking to purchase a gift for someone that you don't like very much, this book is for you.

Terrible Maps emerged on social media in 2015 as a spin-off from Amazing Maps. After posting interesting, sensible, factual maps on social media for a few years as Amazing Maps it became clear that humans inherently like to whine because every comment section was full of complaints, arguments and general negative feedback. I was struck with an idea. Why not post maps that aren't meant to be good, that defy rational criticism, that transcend the boundaries of 'right' or 'wrong'? Terrible Maps was born.

Initially created as nothing more than a joke, the global thirst for terrible maps has been phenomenal. Across Twitter, Instagram and Facebook there are now more than 2.8 million followers of Terrible Maps at the time of writing – no wonder the world seems so lost!

Within these pages, I present to you a condensed assortment of some of the most terrible maps ever conceived. Fasten your seatbelts and grab your compass as we embark on a fun-filled journey that will challenge your notions of scale and sense of direction, and remind you never to take a map too seriously.

Is it against the law to knock on a door and run away?

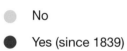

- ○ No
- ● Yes (since 1839)

The top 12 states to live in

Indonesia

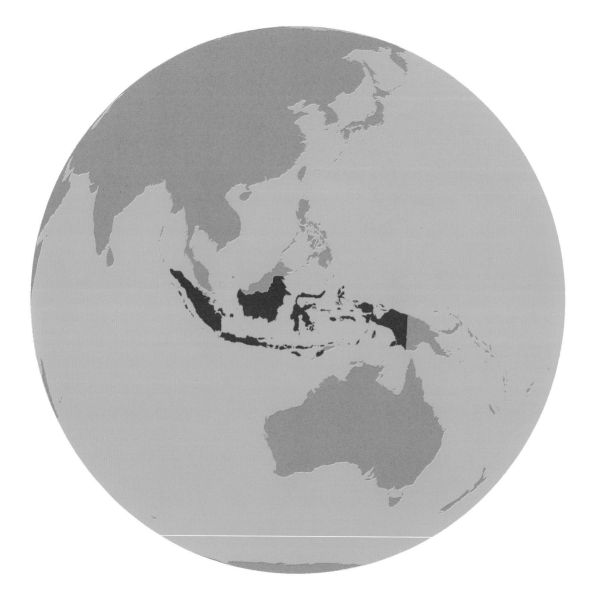

Outdonesia

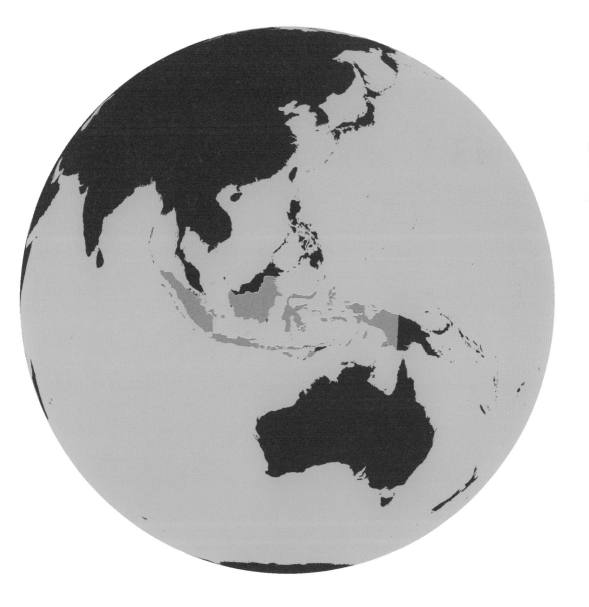

Saudi Arabia mapped only by its rivers

Cat earth theory

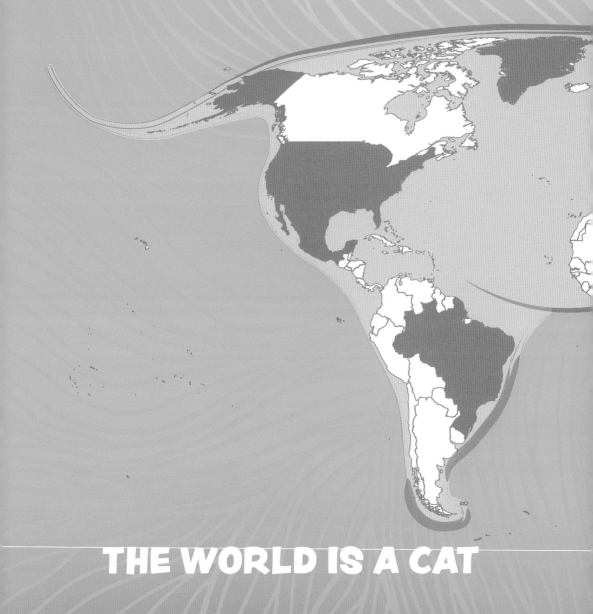

THE WORLD IS A CAT

PLAYING WITH AUSTRALIA

Dog's head

Australia: you'll never see it the same way again

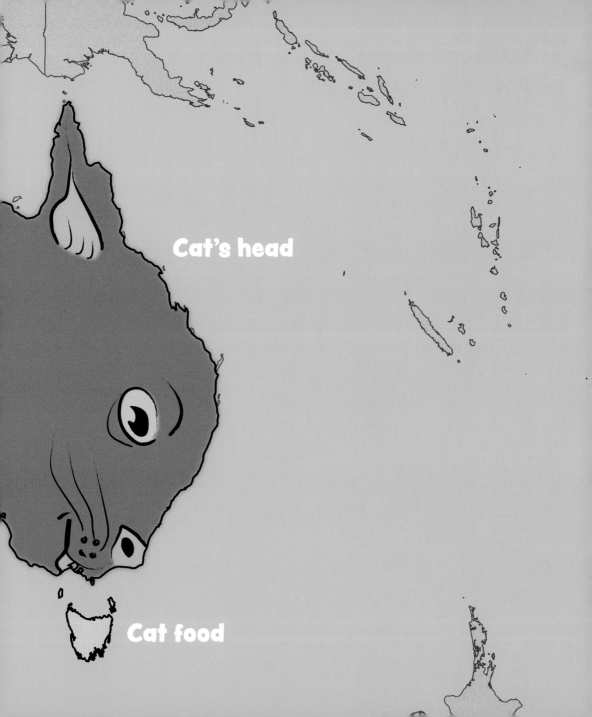

Oklahoma if it was used to express feelings

Oklahoma

Goodlahoma

Badlahoma

The 'international community'

Coldest season of the year

Winter

Spring

Summer

Autumn

NORWAY

Oslo

SWEDE

Stockho

North
Sea

DENMARK

IRELAND

Dublin

Copenhagen

UNITED
KINGDOM

ATLANTIC

OCEAN

London

Amsterdam

NETHERLANDS

Elbe

Berlin

POLA

Brussels

BELGIUM

GERMANY

Oder

Seine

LUXEMBOURG

Prague

Paris

CZECHIA

Loire

Rhin

FRANCE

Bern

Vienna

SLOV

Garonne

LIECHT.

SWITZERLAND

AUSTRIA

Brati

Budapest

PORTUGAL

Duero

Rhône

Zagreb

HUNG

Ebro

ANDORRA

CROATIA

Lisbon

Madrid

BOSNIA &
HERZ.

SPAIN

Corsica

ITALY

Sarajevo

Balearic Islands

Rome

MONTENEGRO

Podgorica

Sardinia

Tirana

Mediterranean Sea

ALBAN

Argentina

Aren'tgentina

Where you can marry your first cousin in the United States

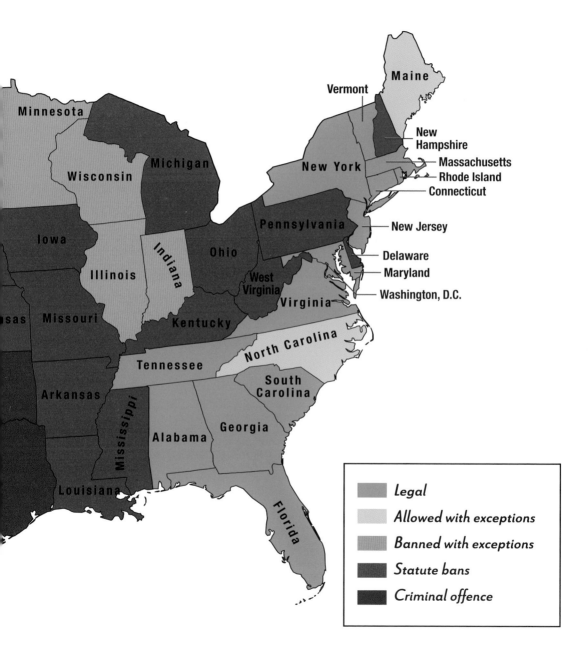

	Legal
	Allowed with exceptions
	Banned with exceptions
	Statute bans
	Criminal offence

Miley Cyprus

Mediterranean Sea

Cape
Kormakitis

Lapithos

Kyrenia

Morphou
Bay

Fama
Ba

Fama

Cape
Arnaoutis

Kokkina

Polis

Larnaka

Peyia

Kiti

Paphos

Zygi

Episkopi

Lemosos

Akrotiri Bay

Akrotiri

Cape Gata

Most common answer from respondents after being asked what state they are in

No data

Map of Roman air bases in 2nd century AD

Coincidence?

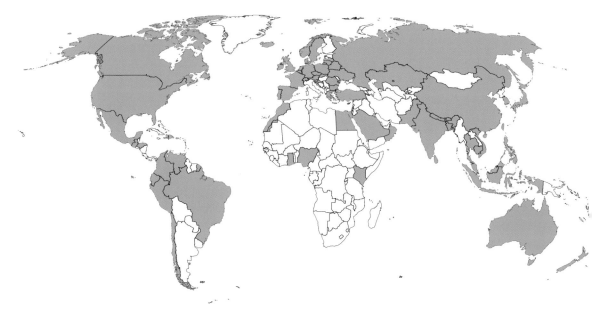

Countries that have Domino's pizza

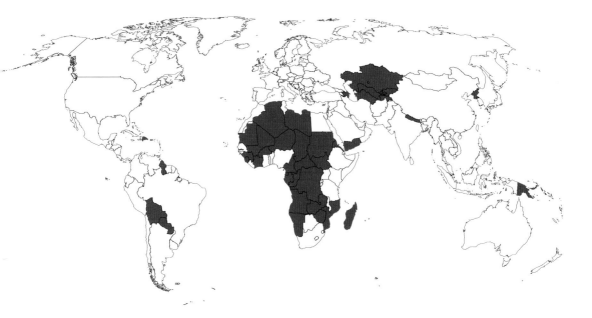

Countries never visited by a sitting US president

Fatalities from texting and driving in 1960

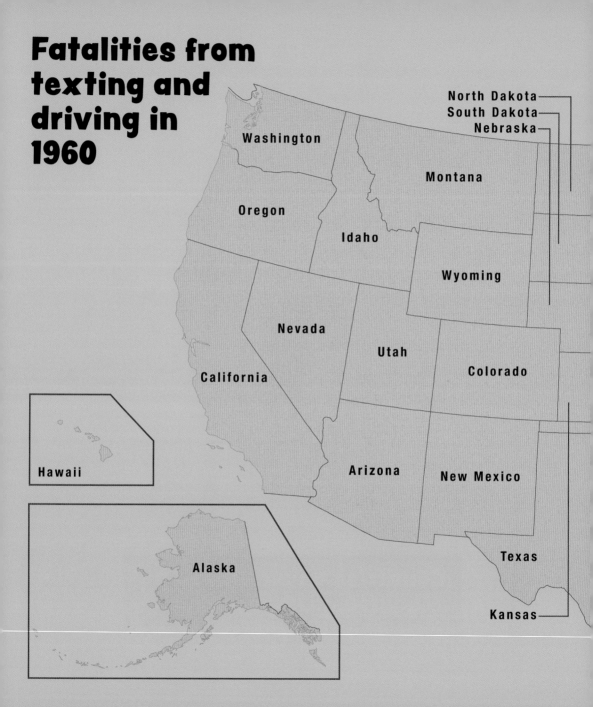

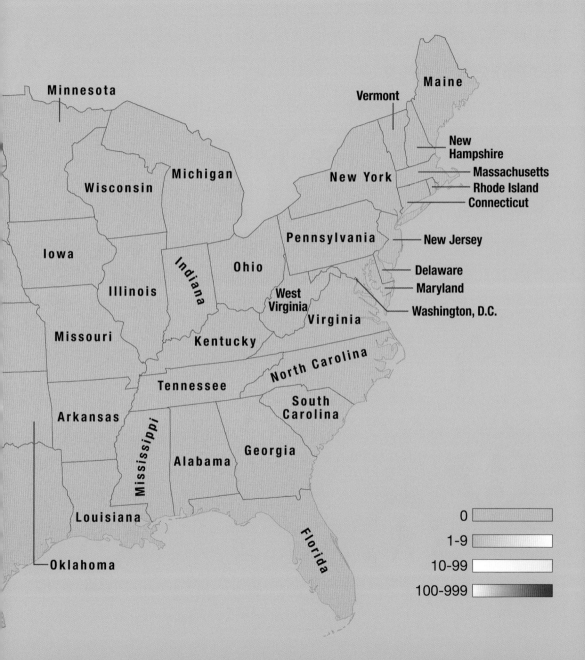

Traffic signs for 'No pedestrians' across Europe

More people live in the green area than the blue area

You can sail in a straight line from England to France

Cayenne
(French Guiana)

Southampton

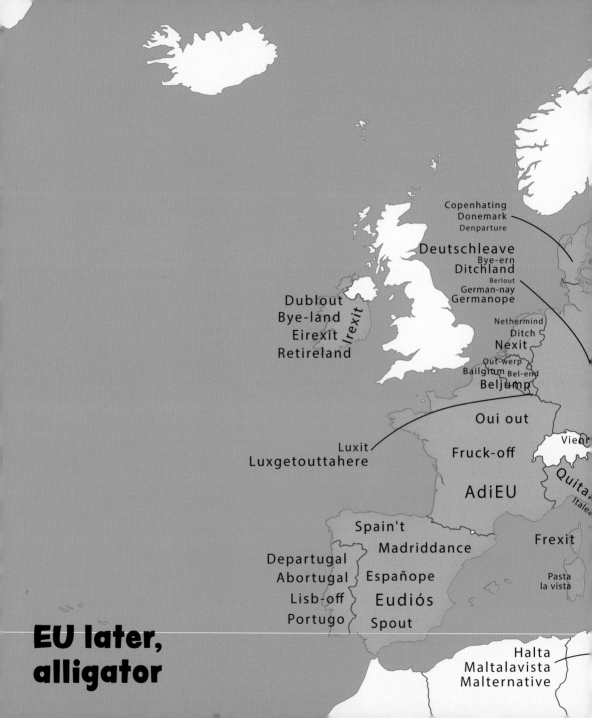

Copenhating
Donemark
Denparture

Deutschleave
Bye-ern
Ditchland
Berlout
German-nay
Germanope

Nethermind
Ditch
Nexit

Dublout
Bye-land
Eirexit
Retireland

Irexit

Out-werp
Bailgium Bel-end
Beljump

Oui out

Luxit
Luxgetouttahere

Fruck-off

AdiEU

Vien

Quita

Itale

Spain't
Madriddance

Frexit

Departugal
Abortugal
Lisb-off
Portugo

Españope
Eudiós
Spout

Pasta
la vista

**EU later,
alligator**

Halta
Maltalavista
Malternative

Map of the United States if it was an Oregon donor

An extensive map of all countries that use the MM/DD/YYYY date format

Yep

Railway map of
Antarctica

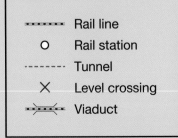

- ┄┄┄ Rail line
- ○ Rail station
- ┈┈┈ Tunnel
- ✕ Level crossing
- ┉┉┉ Viaduct

The great British bread off

Buttery
Aberdeen roll
Rowie
Morning roll

Bun

Muffin
Map
Morning roll

Blackpool burner
Barm cake
Stotty
Scuffler — Tea cake
Barm
Breadcake
Oven bottom
Bun
Muffin
Cob
Batch
Bara

Blaa

Roll

I can't unsee this

Electricity consumption in Europe in 1507

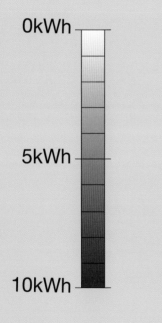

0kWh

5kWh

10kWh

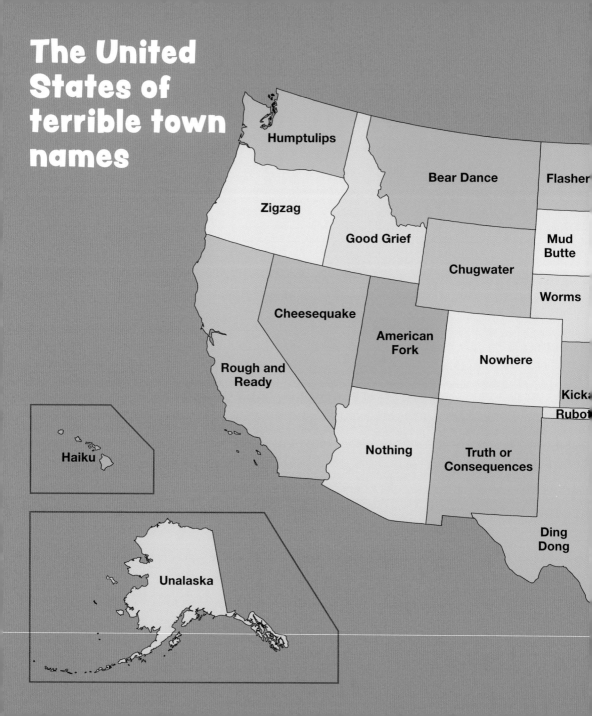

Nowthen

Imalone

Knockemstiff

Bread Loaf

Bald
Head

Happy Corner

Hell

Horseheads

Satan's Kingdom
Scituate
Hazardville

Hard
Scratch

Goofy
Ridge

Santa
Claus

Big Beaver

Buttzville

Lick
Fork

Blue Ball
Boring

Humansville

Monkey's
Eyebrow

Tightsqueeze

Stinking Creek

Boogertown

Ketchup-
town

Bald
Knob

Hot
Coffee

Choc-
colocco

Hopeulikit

Uneedus

Yeehaw
Junction

Portugal is smaller than the United States, Russia and China combined

Map of famous places

Spainsley Harriott

Argentina Turner

Nepal McCartney

Bolivia Coleman

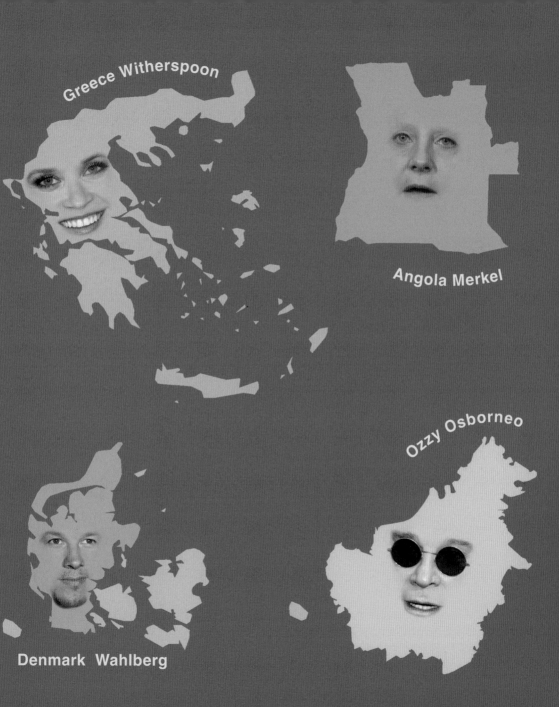

Greece Witherspoon

Angola Merkel

Denmark Wahlberg

Ozzy Osborneo

US states but the first letter is missing

Ashington

Ontana

Orth
Dakota

Regon

Daho

Outh
Dakota

Yoming

Ebraska

Evada

Tah

Olorado

Ans

Alifornia

Klahor

Awaii

Rizona

Ew
Mexico

Exas

Laska

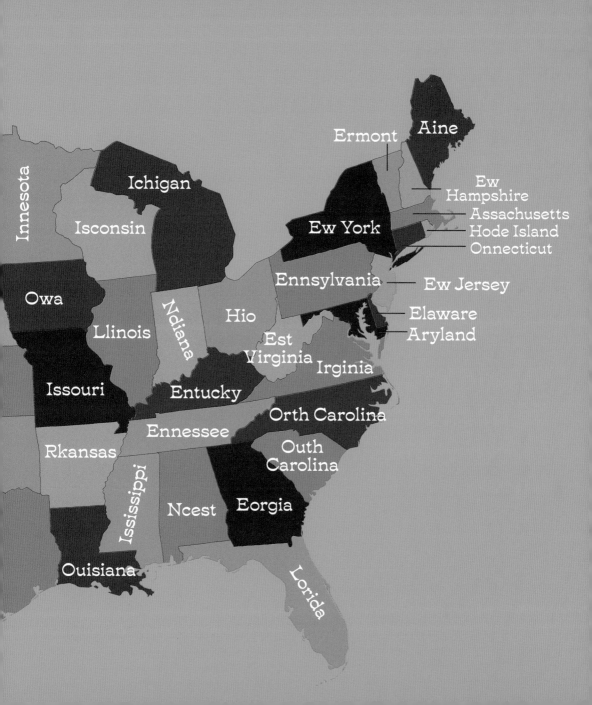

World map
(if approaching
Earth from
below)

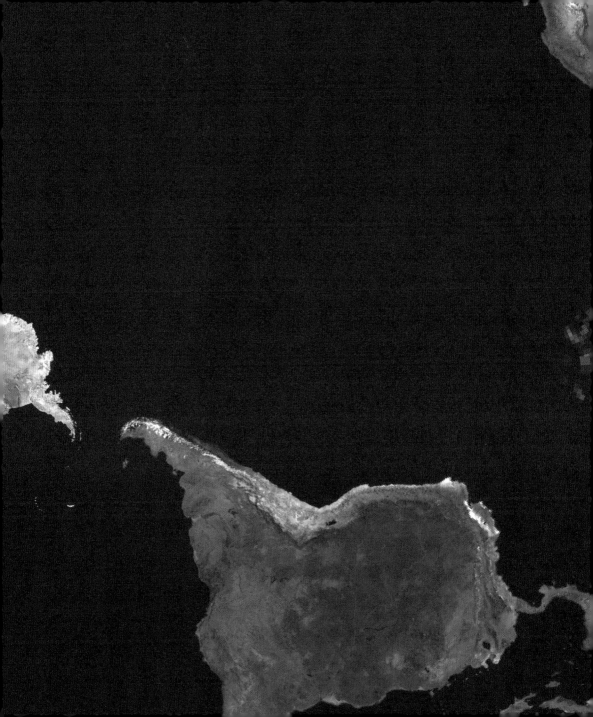

**Countries
arranged by
geographical
location**

Literally all statistics about Italy

Map of every
pub in the UK

Is this country Greenland?

● Yes
● No
● No data
● Nope, that's Mexico

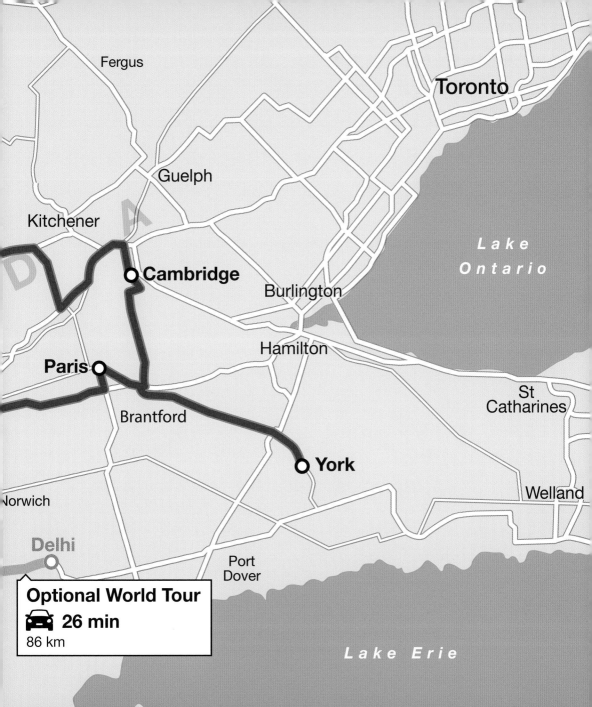

Fergus

Toronto

Guelph

Kitchener

Cambridge

Burlington

*Lake
Ontario*

Hamilton

Paris

St
Catharines

Brantford

York

Norwich

Welland

Delhi

Port
Dover

Optional World Tour
🚗 **26 min**
86 km

Lake Erie

Place names of the UK and Ireland

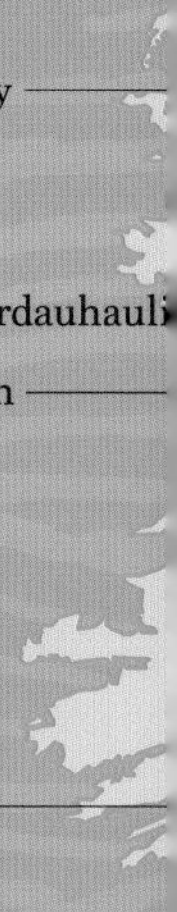

Tubbercurry ——

Muckanaghederdauhaulia

Llanfairpwllgwyngyllgogerychwyrndrobwllllantysiliogogogoch ——

Kilbrittain ——

OCD USA

Moon on flag vs flag on moon

Countries with moon on flag

Countries with flag on moon

Countries

COMA

Words for 'Coma' in European languages (someone didn't get the memo...)

KOM

KOMA

COMA

COMA KOMA

KOMA

COMA

K

COMA

COM

COMA COMA

The most popular word in each state

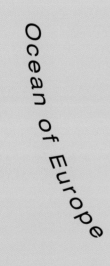

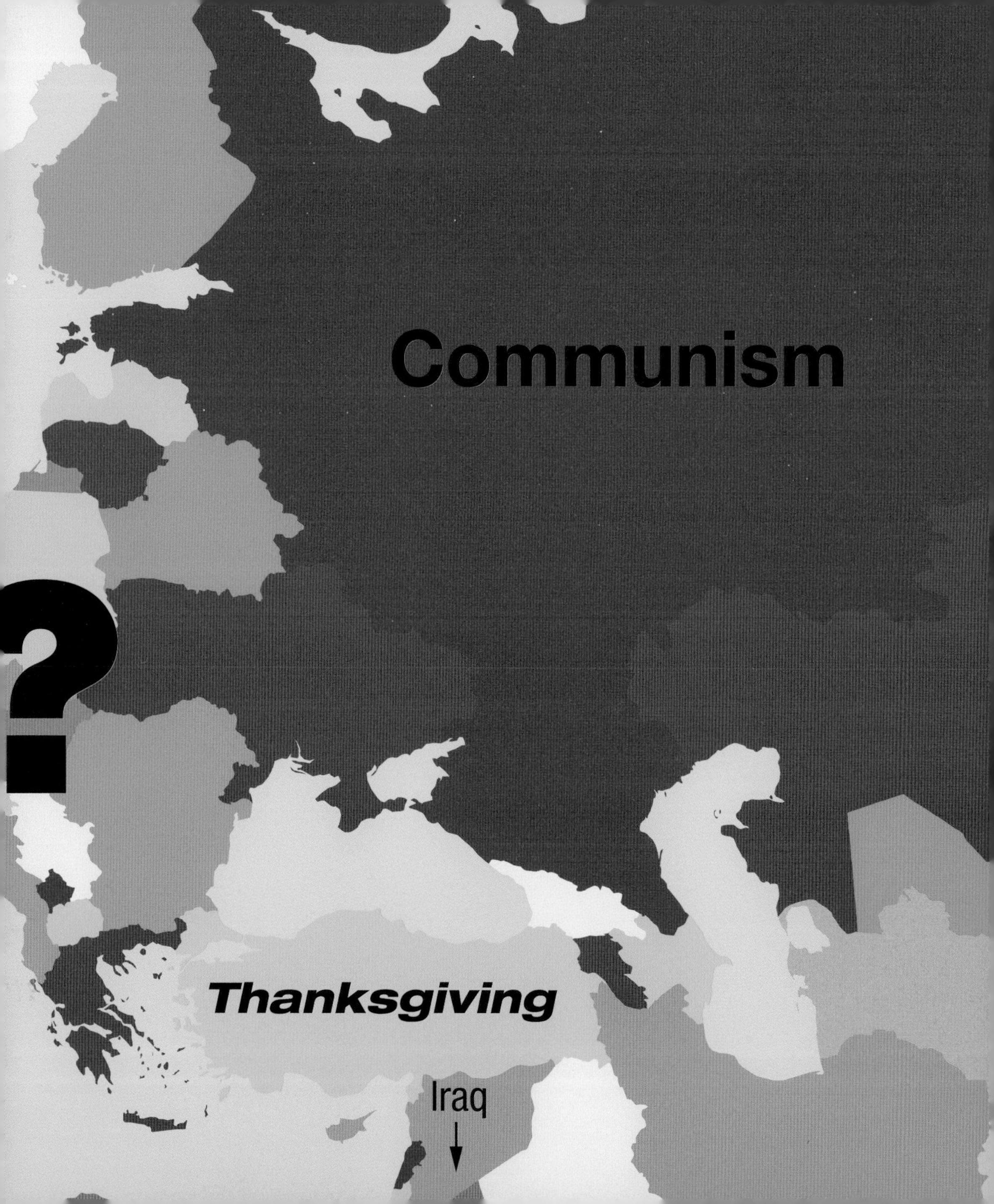
Communism
?
Thanksgiving
Iraq

How aliens see Earth

If Great Britain was located next to Japan

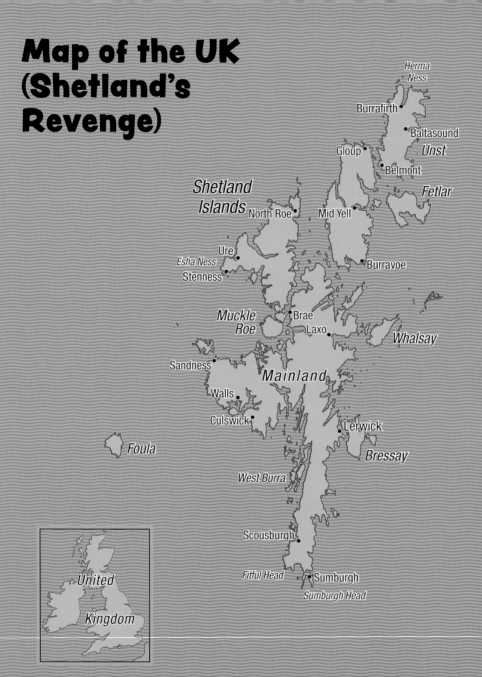

Map of the UK (Shetland's Revenge)

Herma Ness

Burrafirth

Baltasound

Gloup

Unst

Belmont

Shetland Islands

Fetlar

North Roe

Mid Yell

Ure

Esha Ness

Burravoe

Stenness

Muckle Roe

Brae

Laxo

Whalsay

Sandness

Mainland

Walls

Culswick

Lerwick

Foula

Bressay

West Burra

Scousburgh

Fitful Head

Sumburgh

Sumburgh Head

United Kingdom

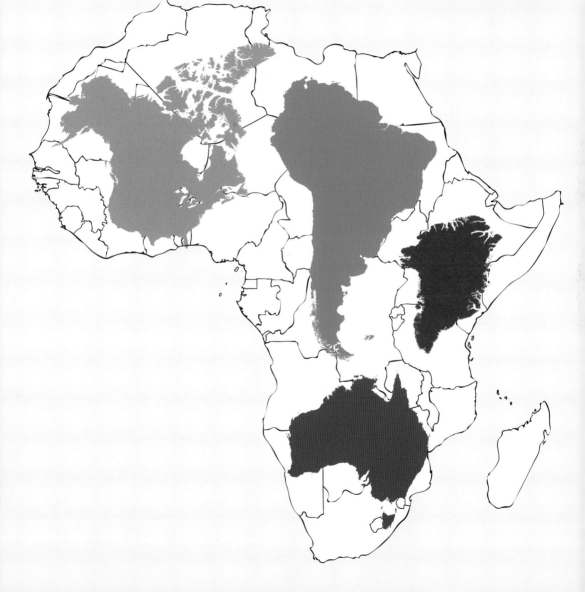

The true size of Africa

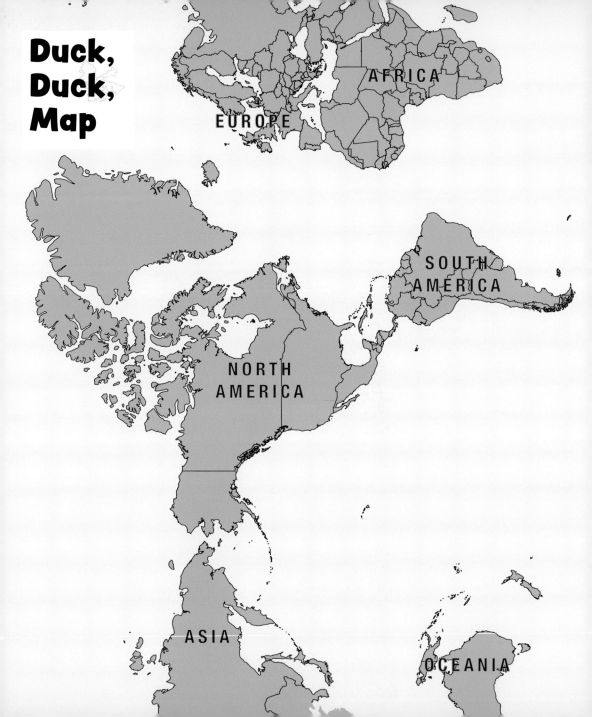

Duck, Duck, Map

EUROPE

AFRICA

SOUTH AMERICA

NORTH AMERICA

ASIA

OCEANIA

More people live inside the circle than outside of it

Now we know what they get up to in Finland...

Public saunas in Finland

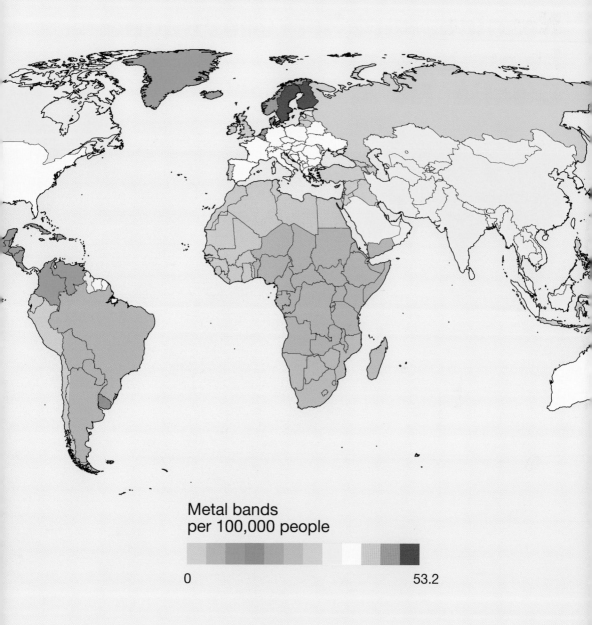

Metal bands
per 100,000 people

0   53.2

Bosnians and Herzegovinians: I wanna swim
Croatia: No

Map of countries that are often omitted from maps

Forgotten

Remembered

Europe but if Europe colonised it

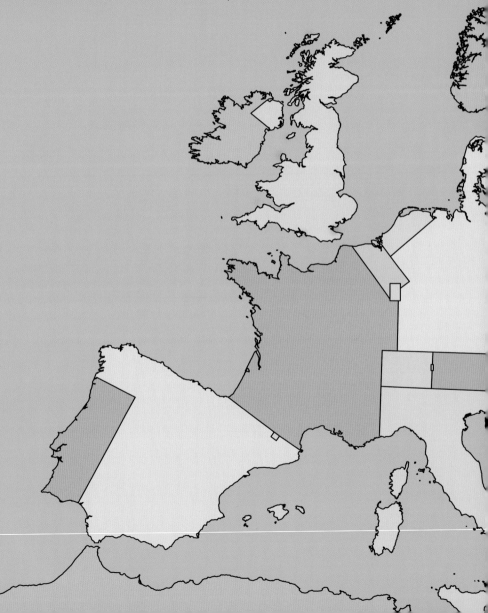

Restaurant order preferences in Africa

FOR HERE

TOGO

Map of Nice people

Nice people   Not Nice people

Prevalence of colour blindness worldwide

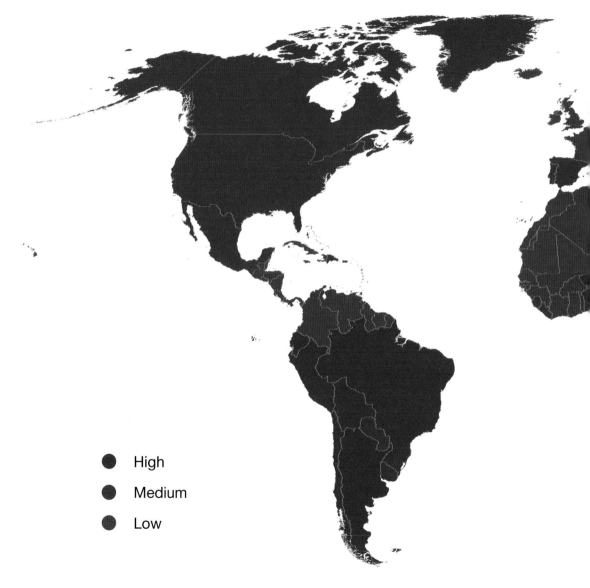

High

Medium

Low

A map of Europe with the naughty bits censored

How to find
Kentucky

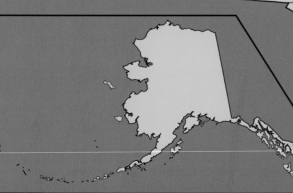

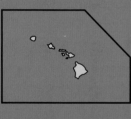

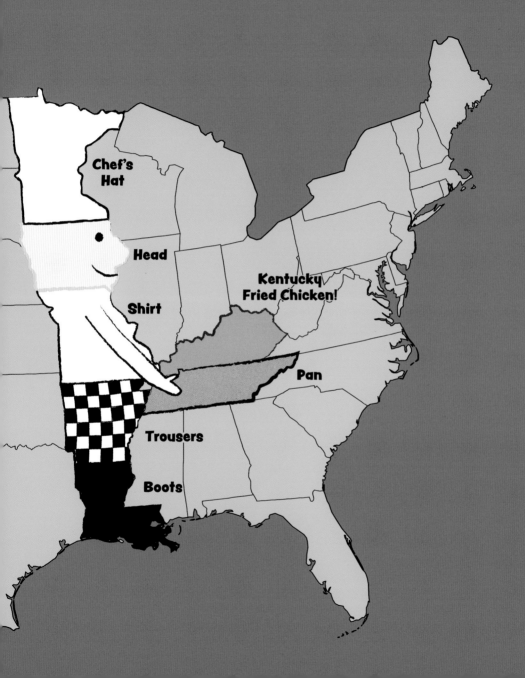

Countries that eat their national animals

Swan
Dragon
European Rabbit
Bull

Bison

Yummy

No

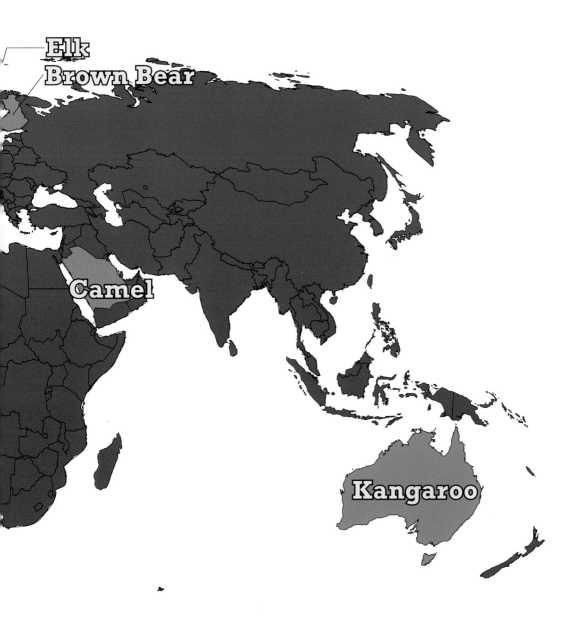

How it started

How it's going

South America vs Pluto

A Southerner's view of the UK and Ireland

Scandinavia

The Far North

Whisky
Fried Mars bars

Golf
The nice place in Scotland

The old North/ South divide

Some rocks

George Best

Walk here... good for holidays even though up North

Ant and Dec

The North

Guinness

The Beatles

Footballers live here

Dragons... and anxious sheep

Silly accents

Farmers with six fingers

Oxford

Cambridge

Stag dos

Farmers

Fake tan

The South

Holiday cottages

Cider

Retired people

Pirates

Map of Earth if there was no land

GREAT PACIFIC GARBAGE PATCH

Average flag colour
by latitude

World map after an 8,800m sea level rise

*Everest
Island*

Maposaurus rex

The legality of waking a sleeping bear to take its picture

Vermont Maine
nnesota
Michigan
New Hampshire
Massachusetts
Wisconsin
New York
Rhode Island
Connecticut
Pennsylvania
New Jersey
owa
Ohio
Delaware
Illinois
Indiana
West
Maryland
Virginia
Washington, D.C.
Missouri
Kentucky
Virginia
North Carolina
Tennessee
Arkansas
South
Carolina
Mississippi
Georgia
Alabama
Louisiana
Florida

Legal
Illegal

Super Bowl wins by country

Super Bowl wins by country

0   28   57

No data

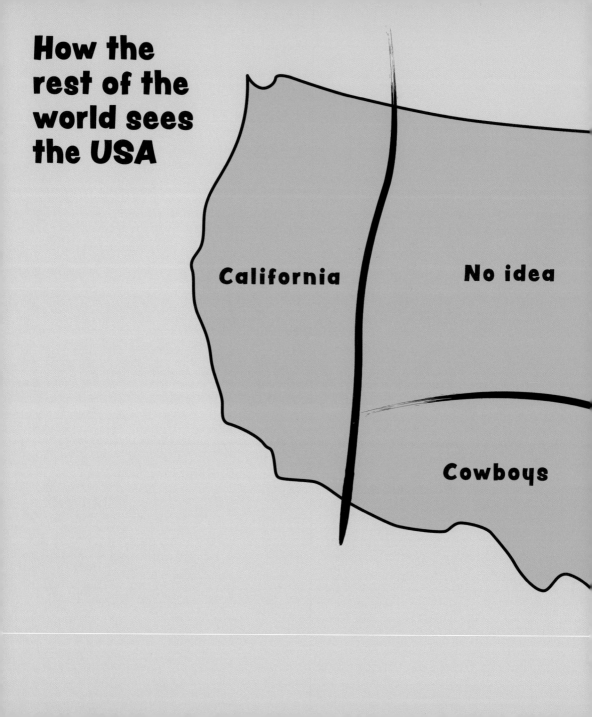

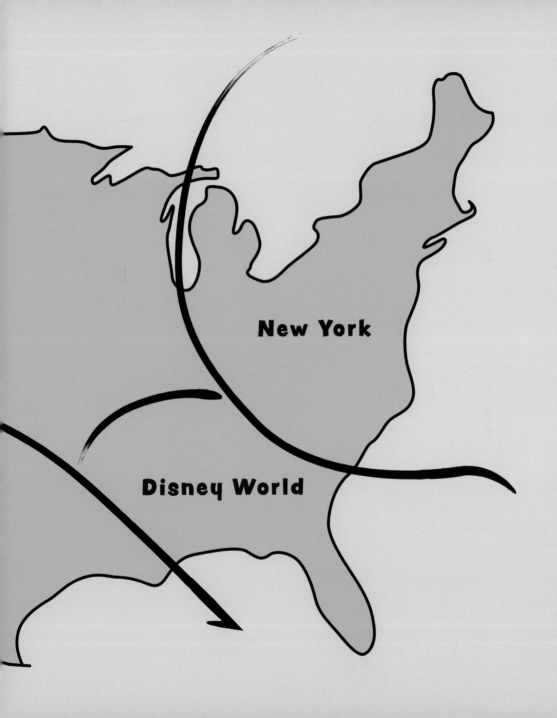

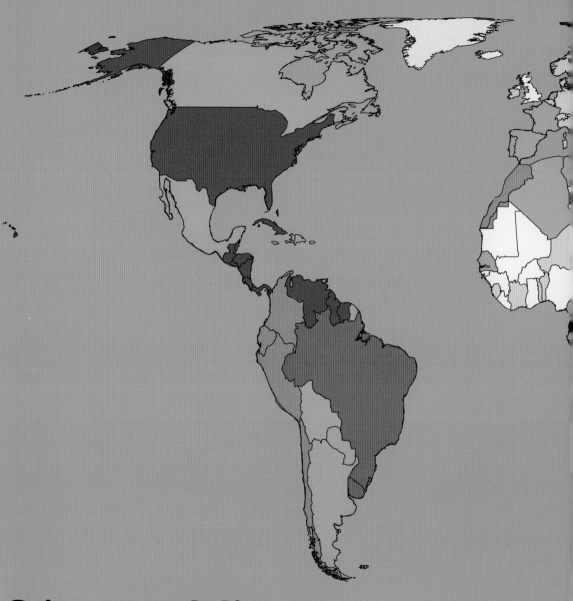

**Prison population per
100,000 people**

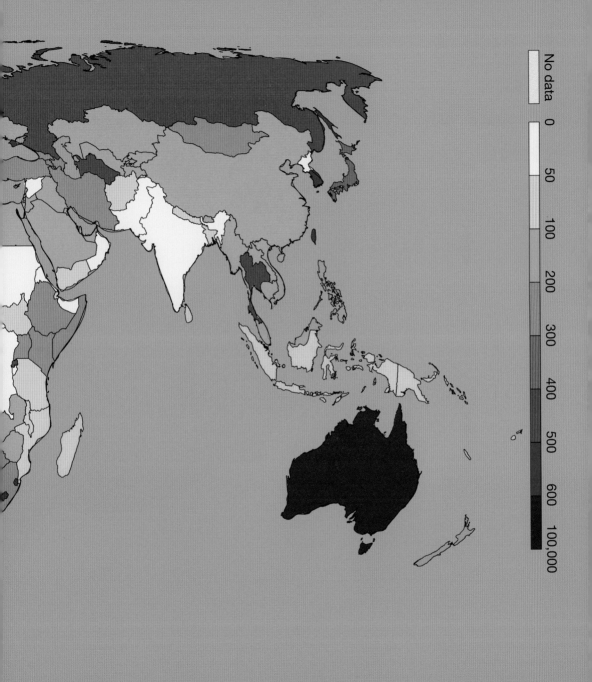

Average jeans colour by state

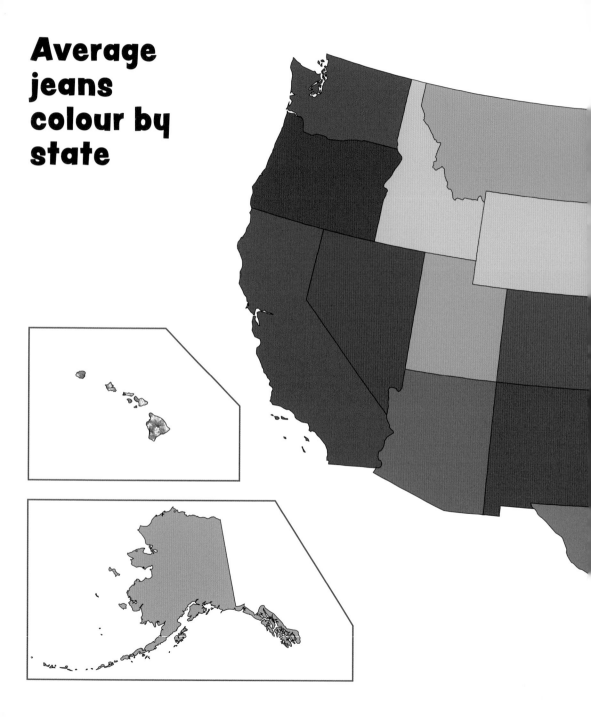

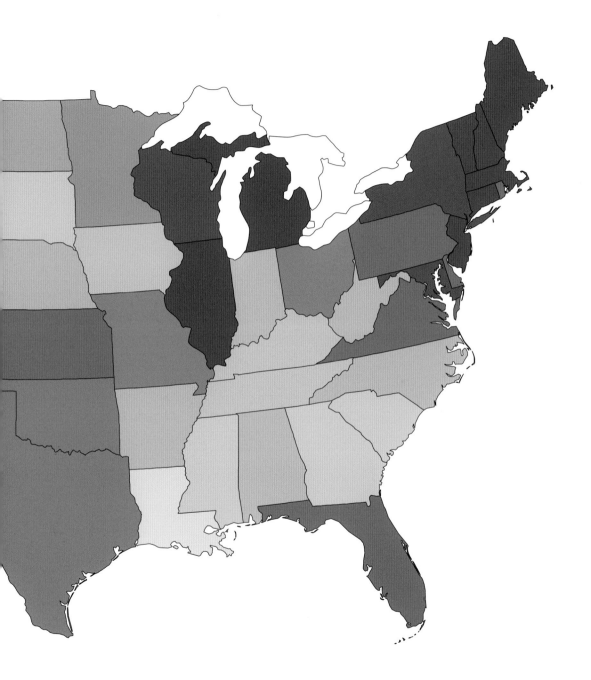

Map of South America

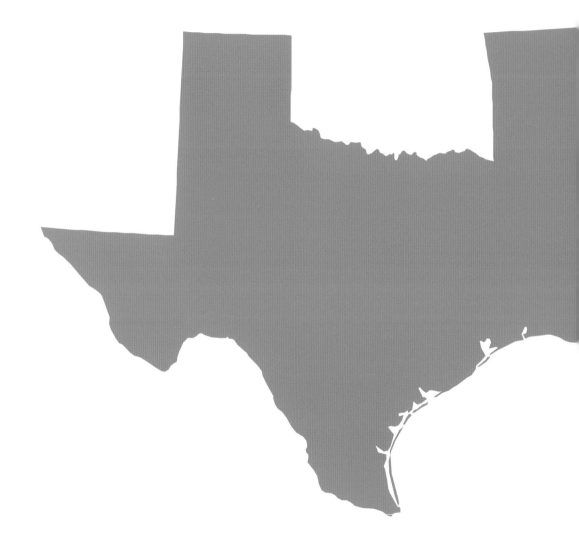

Which way do you screw?

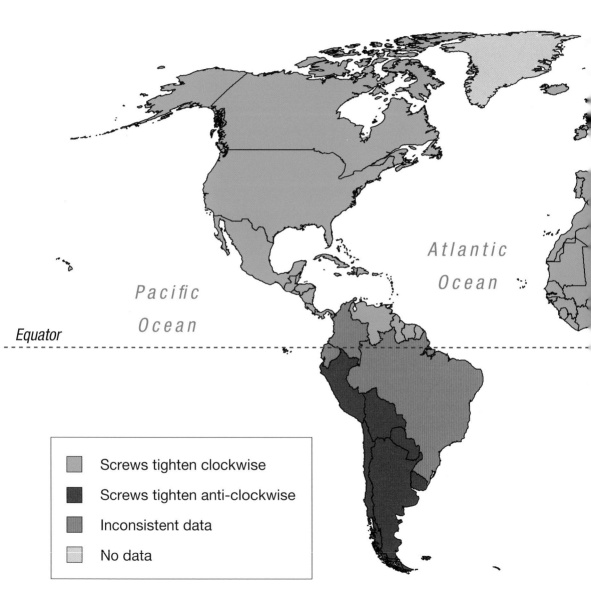

Atlantic
Ocean

Pacific
Ocean

Equator

Screws tighten clockwise

Screws tighten anti-clockwise

Inconsistent data

No data

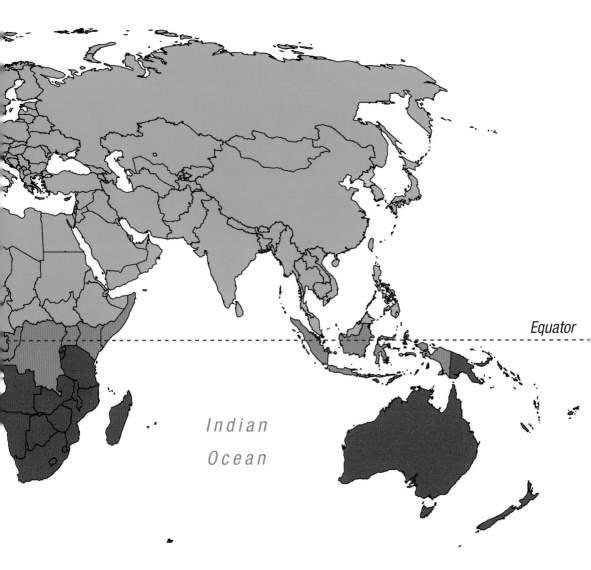

Equator

Indian

Ocean

Countries that have declared war on birds ... and lost

Countries that have not
declared war on birds

Countries that have
declared war on birds

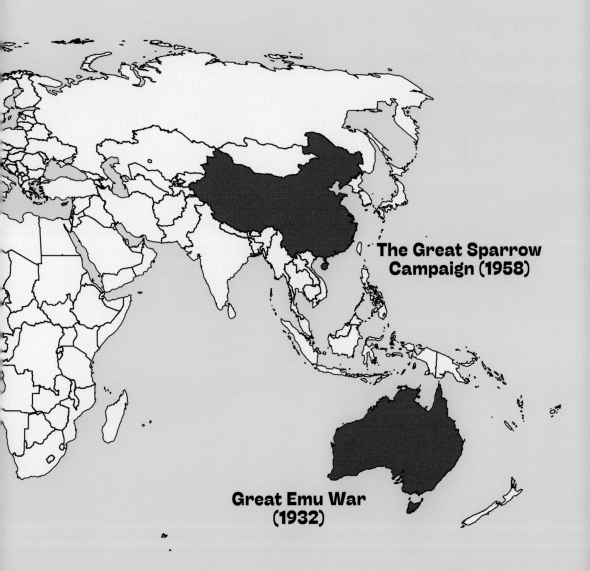

The Great Sparrow Campaign (1958)

Great Emu War (1932)

The most widely spoken language other than Spanish by state

E n g l

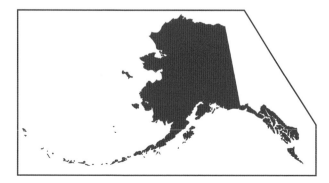

Countries by human development if human development was based on the number of wild penguins

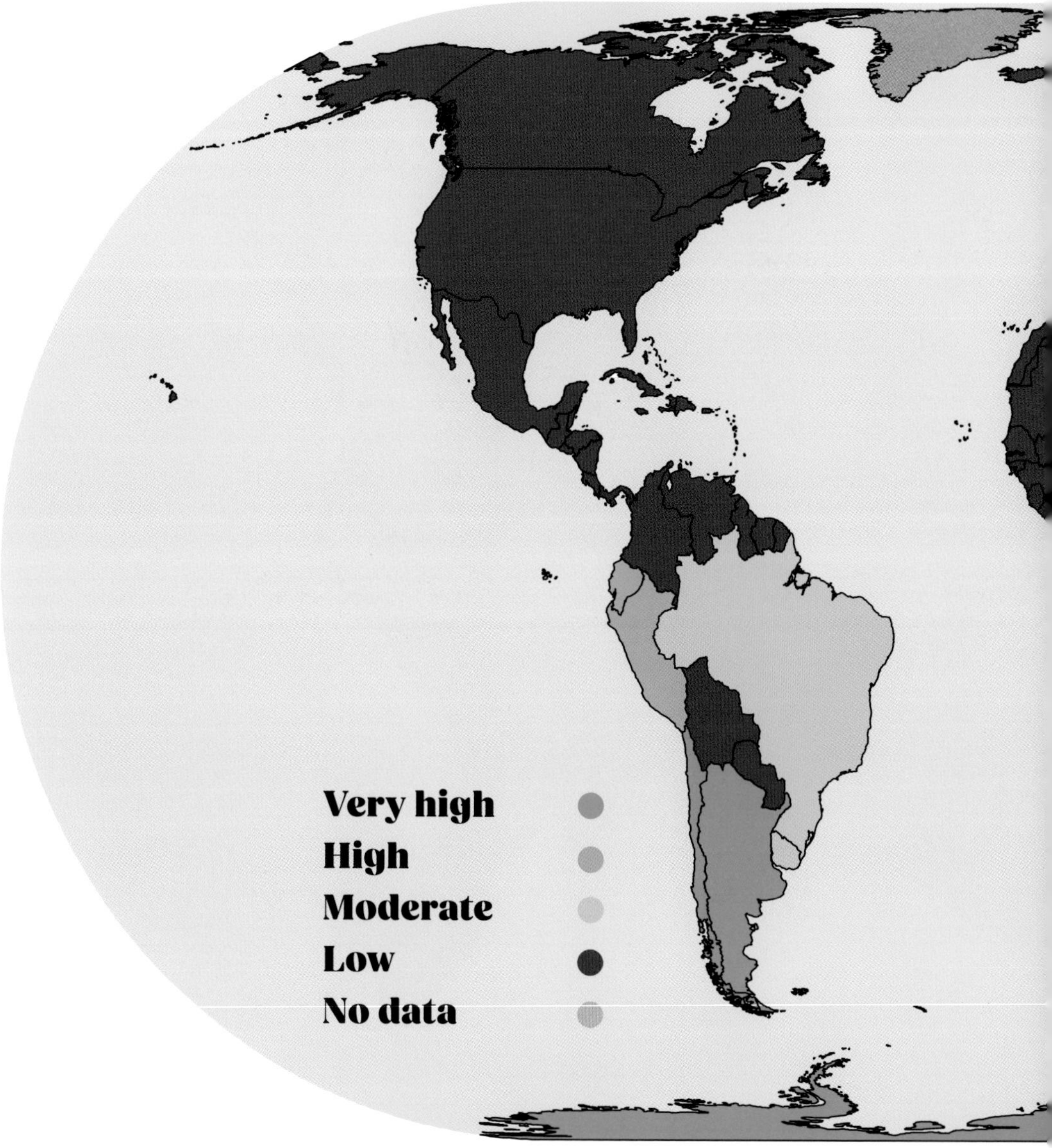

Very high
High
Moderate
Low
No data

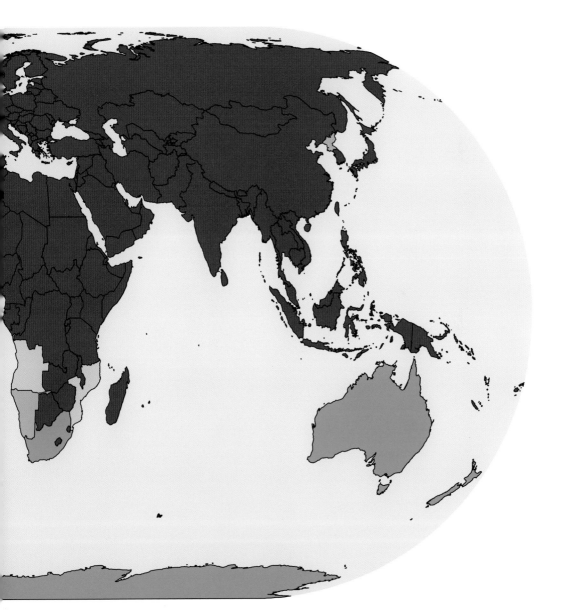

Map showing the origin
of Miss Universe winners
(1952-present)

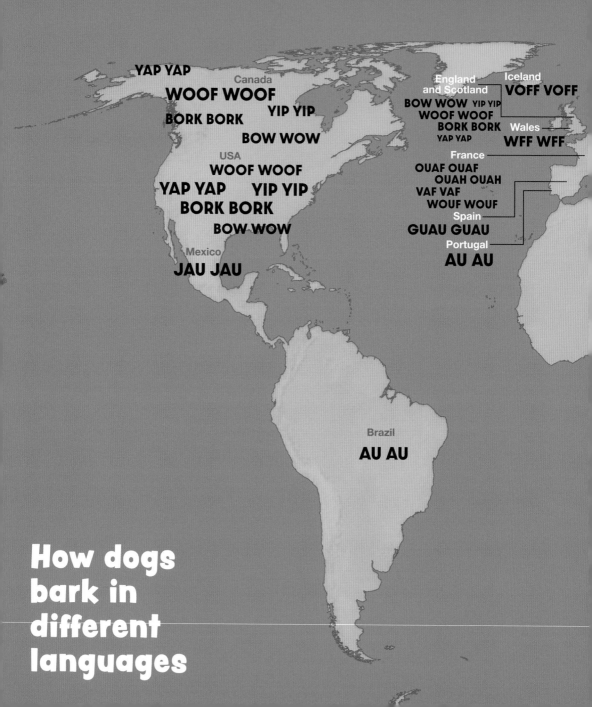

How dogs bark in different languages

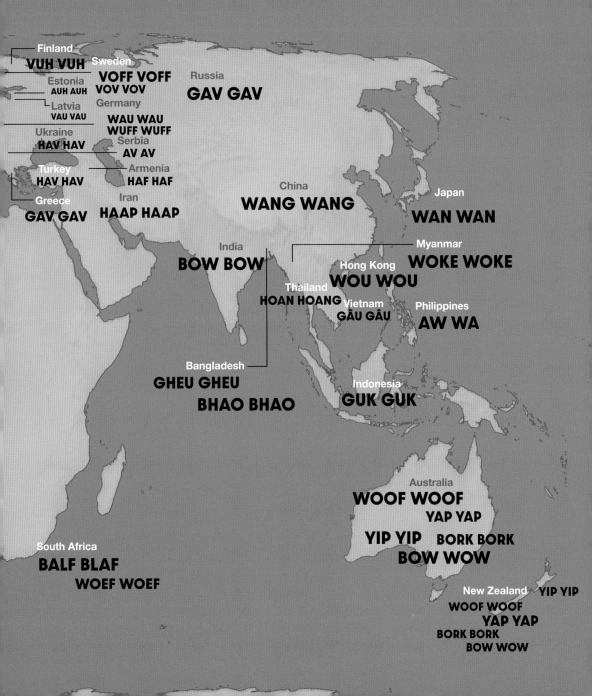

Finland
VUH VUH

Sweden
VOFF VOFF
VOV VOV

Estonia
AUH AUH

Latvia
VAU VAU

Germany
WAU WAU
WUFF WUFF

Ukraine
HAV HAV

Serbia
AV AV

Russia
GAV GAV

Turkey
HAV HAV

Armenia
HAF HAF

Greece
GAV GAV

Iran
HAAP HAAP

China
WANG WANG

Japan
WAN WAN

India
BOW BOW

Myanmar
WOKE WOKE

Hong Kong
WOU WOU

Thailand
HOAN HOANG

Vietnam
GÂU GÂU

Philippines
AW WA

Bangladesh
GHEU GHEU
BHAO BHAO

Indonesia
GUK GUK

Australia
WOOF WOOF
YAP YAP
YIP YIP   **BORK BORK**
BOW WOW

South Africa
BALF BLAF
WOEF WOEF

New Zealand   **YIP YIP**
WOOF WOOF
YAP YAP
BORK BORK
BOW WOW

Data!

Does Terrible Maps have a larger social media following than the population of your country?

Yes

No

No but in light green

HarperCollins*Publishers*
1 London Bridge Street
London SE1 9GF

www.harpercollins.co.uk

HarperCollins*Publishers*
Macken House, 39/40 Mayor Street Upper
Dublin 1, D01 C9W8, Ireland

First published by HarperCollins*Publishers* **2023**

10 9 8 7 6 5 4

Text © Terrible Maps 2023
Map design © Lovell Johns Ltd 2023

Terrible Maps asserts the moral right to be identified as the author of this work

A catalogue record of this book is available from the British Library

ISBN 978-0-00-864159-7

Printed and bound in Latvia

MIX
Paper | Supporting responsible forestry
FSC™ C007454

This book is produced from independently certified FSC™ paper to ensure responsible forest management.

For more information visit: www.harpercollins.co.uk/green

Additional interior images: Shutterstock.com

with the exception of:
Miley Cyprus: Jon Kopaloff / FilmMagic / Getty Images,
I can't unsee this: Artefact / Alamy Stock photo,
Map of famous places: (Tina Turner) Tim Mosenfelder / Getty Images,
(Reese Witherspoon) Steve Granitz / FilmMagic / Getty Images

If you have any terrible maps to suggest you can
connect with us on:

🐦 @TerribleMaps

📷 @TerribleMap

📘 Terrible Maps